THE POETRY OF NEON

The Poetry of Neon

Walter the Educator™

SKB

Silent King Books a WhichHead Imprint

Copyright © 2023 by Walter the Educator™

All rights reserved. No part of this book may be reproduced in any manner whatsoever without written permission except in the case of brief quotations embodied in critical articles and reviews.

First Printing, 2023

Disclaimer
This book is a literary work; poems are not about specific persons, locations, situations, and/or circumstances unless mentioned in a historical context. This book is for entertainment and informational purposes only. The author and publisher offer this information without warranties expressed or implied. No matter the grounds, neither the author nor the publisher will be accountable for any losses, injuries, or other damages caused by the reader's use of this book. The use of this book acknowledges an understanding and acceptance of this disclaimer.

"Earning a degree in chemistry changed my life!"
- Walter the Educator

dedicated to all the chemistry lovers, like myself, across the world

CONTENTS

Dedication v

Why I Created This Book? 1

One - Star Of The Show 2

Two - Dreams To Elope 4

Three - Light Shine 6

Four - More Than A Gas 8

Five - Neon's Brilliance 10

Six - Reign Supreme 12

Seven - Palette So Bold 14

Eight - Masterpiece Within The Dark 16

Nine - Forever To Last 18

Ten - Dazzling Scene 20

Eleven - Element Grand 22

Twelve - Captivate The Senses 24

Thirteen - Soul Of The Night	. . . . .	26
Fourteen - Element Bright	. . . . .	28
Fifteen - Darkness Retreats	. . . . .	30
Sixteen - Kaleidoscope	. . . . .	32
Seventeen - Dance Of Electrons	. . . .	34
Eighteen - Vibrant Hue	. . . .	36
Nineteen - Radiant Spirit	. . . . . .	38
Twenty - Essence Of Beauty	. . . .	40
Twenty-One - Show Us The Way	. . . . . .	42
Twenty-Two - Vivid Display	. . . . . .	44
Twenty-Three - Illuminates The Soul	. . . . .	46
Twenty-Four - Glowing Mass	. . . . .	48
Twenty-Five - Story Of Science	. . . .	50
Twenty-Six - Brighter Day	. . . . . .	51
Twenty-Seven - Magic Of The Atmosphere	. .	53
Twenty-Eight - Captivating And Grand	. . .	55
Twenty-Nine - Radiant Glow	. . . .	57
Thirty - Mysterious Ways	. . . .	59
Thirty-One - Luminescent Dream	. . . . . .	61
Thirty-Two - Hearts Of Creators	. . . . .	63

Thirty-Three - Turning Page	. . . .	65
Thirty-Four - Celestial Light	. . . .	66
Thirty-Five - City Streets To Open Skies	. . .	68
Thirty-Six - Progress And Might	. . . .	70
Thirty-Seven - Paints The Night	. . . .	72
About The Author	. . . .	74

WHY I CREATED THIS BOOK?

Creating a poetry book about the chemical element of Neon allows for a unique and creative exploration of its properties and significance. Poetry has the power to evoke emotions and capture the essence of Neon in a way that scientific texts cannot. By combining science with art, this poetry book about Neon can educate and inspire readers about the wonders of science, while also highlighting the beauty in seemingly mundane elements. It provides an opportunity to delve into the ethereal glow, the electrifying properties, and the captivating symbolism of Neon, making it a fascinating subject for poetic expression.

ONE

STAR OF THE SHOW

In the realm of atomic light,
A shimmering brilliance takes flight.
Neon, the luminescent star,
A symbol of the heavens afar.

With electrons dancing, oh so bright,
In the outer shell, a captivating sight.
A noble gas, it proudly stands,
Unreactive, in noble strands.

In tubes and signs, it comes alive,
A vibrant glow, it does contrive.
A kaleidoscope of colors true,
Neon's radiance breaks through.

In the night, a neon haze,
Guiding lost souls through the maze.
A city's heartbeat, a pulsing beat,
Neon lights, an urban retreat.

From Las Vegas to Tokyo's streets,
Neon's magic, the eye it meets.
A symphony of vibrant hues,
A spectacle that cannot lose.

Oh, Neon, you fill the night,
With your electrifying light.
A symbol of energy, bold and free,
A testament to possibility.

So, let us revel in your glow,
Neon, the star of the show.
A beacon in the darkness deep,
A reminder that dreams we can keep.

TWO

DREAMS TO ELOPE

In the realm of celestial fire,
Where stars dance and dreams aspire,
A noble gas, vibrant and bright,
Neon, a beacon in the night.

An element, so rare and grand,
A symbol of heavens, where dreams expand,
Its atomic number, ten, so small,
Yet it shines the brightest of them all.

Unreactive, noble, and true,
Neon, a gas with a vibrant hue,
Its electrons, tightly bound in embrace,
Radiate energy with captivating grace.

In the cityscape's nocturnal glow,
Neon's light puts on a dazzling show,
Signs and billboards, a vibrant sight,
Guiding lost souls through the darkest night.

In bars and cafes, streets alive,
Neon signs, they vividly thrive,
Their luminous glow, a siren's call,
Drawing wanderers, one and all.

Neon's brilliance, an electric storm,
Igniting passions, dreams take form,
A symphony of colors, a flickering dance,
Transcending reality with a luminous trance.

In the depths of night, where shadows creep,
Neon's glow promises secrets to keep,
A whispered promise, a tale untold,
Of energy, possibility, and dreams to unfold.

Oh, Neon, you guide us through the haze,
Illuminate paths in mysterious ways,
With your radiant light, you inspire,
Fulfilling dreams, setting hearts afire.

In this vast universe, so vast and wide,
Neon, you shimmer with cosmic pride,
A symbol of energy, brilliance, and hope,
A celestial jewel, with dreams to elope.

THREE

LIGHT SHINE

Neon, a shimmering brilliance in the night,
A gas that fills the air with its pure light.
In signs and cityscapes, you brightly glow,
A beacon of colors, a mesmerizing show.

Your noble gas nature, so calm and serene,
Inert and stable, a constant light unseen.
Through tubes and bulbs, your energy flows,
Creating a spectacle that everyone knows.

Oh Neon, you guide the lost souls,
In the darkest nights, you make them whole.
Your luminous presence, a celestial guide,
Leading us through darkness, side by side.

You ignite passions, like a fiery flame,
In hearts and minds, you leave your name.
With your vibrant hues, you paint the sky,
A symphony of colors, as dreams come alive.

Neon, you're more than just a gas,
A muse for artists, a symbol of the past.
In the depths of science, you hold the key,
To understanding the universe, for all to see.

So let your light shine, Neon, bright and strong,
A testament to beauty, a cosmic song.
In the vast expanse, you'll forever gleam,
An element of wonder, in our wildest dreams.

FOUR

MORE THAN A GAS

    Neon, a shimmering light, so bright,
In the darkness, you ignite,
A vibrant glow, a neon sign,
Guiding us through the urban grind.
    In the city's bustling streets,
Your luminous colors meet,
Advertising dreams and desires,
A spectacle that never tires.
    Neon, you dance with the night,
Illuminating with all your might,
A kaleidoscope of hues so bold,
You make our stories unfold.
    From the bars to the theaters grand,
You lend a touch of magic, hand in hand,
Creating a world of endless delight,
A beacon in the darkest night.

Neon, you are more than a gas,
A symbol of our urban mass,
A testament to human dreams,
In your glow, our hope beams.

FIVE

NEON'S BRILLIANCE

In the city's heart, Neon's light does dance,
A vibrant glow, a mesmerizing trance.
Amidst the chaos, it guides lost souls,
Through the concrete jungle, its beauty unfolds.

Electric hues, like fireflies in the night,
Neon signs flicker, casting a bewitching sight.
A glowing oasis in the urban sprawl,
Neon's radiance, an enchanting call.

It adorns the streets with a luminous grace,
A kaleidoscope of colors, an electric embrace.
Through darkened alleys and bustling crowds,
Neon's vibrant glow, a beacon that shrouds.

In the silence of night, it creates a symphony,
A neon orchestra, a vibrant harmony.
Lost in its brilliance, we find our way,
Neon's allure, a guide to a brighter day.

So let Neon's light illuminate your dreams,
In its radiant glow, nothing is as it seems.
A chemical element, both magic and art,
Neon's brilliance, forever etched in the heart.

SIX

REIGN SUPREME

In the realm of lights that dance and gleam,
There exists a noble element, a radiant dream.
Neon, the luminescent star of the night,
A spectacle of colors, a radiant delight.

In tubes of glass, it comes to life,
With electric currents, it shines so bright.
A captivating allure, impossible to ignore,
Neon, the creator of a dazzling decor.

Its presence felt in the city streets,
Neon signs, a symphony that beats.
Guiding souls through the urban night,
With its vibrant glow, a beacon of light.

In dark alleys and crowded squares,
Neon whispers secrets, it silently shares.
Inspiring dreams and igniting desire,
Neon, the muse that never tires.

Its glow paints stories on the walls,
A palette of colors, a magical sprawl.
In neon-lit nights, the world transforms,
As reality and fantasy effortlessly conform.

　　From the smallest cafes to the grandest halls,
Neon's allure enchants, enthralls.
Creating an atmosphere so vivid and grand,
Neon, the artist's brush in the creator's hand.

　　In science and art, it finds its place,
A symbol of brilliance, a visual embrace.
From laboratories to galleries so fine,
Neon's radiance continues to shine.

　　Oh, Neon, you captivate with your mystic hue,
A testament to the wonders science can do.
In the realm of lights, you reign supreme,
An element of beauty, a luminescent dream.

SEVEN

PALETTE SO BOLD

In the realm of atoms, a noble gas shines,
Neon, the element, with brilliance refined.
No celestial guide, but a beacon of light,
A luminescent presence, both day and night.

Its electrons, stable, in a shell so bright,
A dance of energy, a captivating sight.
Neon, the radiant, in colors it glows,
A kaleidoscope of hues, a vibrant show.

In urban landscapes, where darkness prevails,
Neon takes its place, its brilliance prevails.
Signs and billboards, a cityscape's attire,
Neon's vibrant colors set the streets on fire.

A playground for artists, a palette so bold,
Neon paints the night, in stories untold.
From bustling streets to quiet alleyways,
Neon's vivid glow, a mesmerizing display.

So let us celebrate this element rare,
Neon's radiance, beyond compare.
A symbol of beauty, a shining star,
In its luminous presence, we're never far.

EIGHT

MASTERPIECE WITHIN THE DARK

    Neon, the luminescent star,
A radiant glow from afar.
In the darkness it ignites,
Colors dancing through the night.
    Iridescent hues so bold,
A spectacle to behold.
Neon signs light up the street,
Urban landscape, vibrant and complete.
    In cafes and bars it gleams,
Casting neon dreams.
A beacon in the city's heart,
Guiding wanderers through the dark.
    Neon's glow, a modern art,
Captivating right from the start.

With every shade and every hue,
It paints a world that's bold and true.
    So let Neon's light inspire,
Fuel the flames of your desire.
In its brilliance, find your way,
And let your creativity sway.
    Neon, a symbol of the night,
A testament to our delight.
In its glow, we find our spark,
A masterpiece within the dark.

NINE

FOREVER TO LAST

In the heart of the city, Neon awakes,
A vibrant symphony, the night it makes.
Electric dreams, in colors that gleam,
A kaleidoscope of art and esteem.

Neon signs, like stars on the street,
Guiding lost souls with their luminous beat.
A beacon of hope in the darkest of nights,
Igniting the spirit, igniting the sights.

In a world of monotony, Neon's the muse,
A palette of hues, a canvas to choose.
From humble beginnings, it soars up high,
Captivating hearts with its radiant sky.

An element of wonder, it dances and glows,
A spectacle of beauty, the whole world shows.
With each flicker and hum, it tells a tale,
Of dreams and desires, it will never fail.

So let us bask in its ethereal glow,
Embrace the enchantment, let our spirits grow.
For Neon is more than just a mere gas,
It's a symbol of dreams, forever to last.

TEN

DAZZLING SCENE

In cities filled with electric glow,
Where vibrant colors start to show,
There lies a shimmering, neon light,
Casting its magic in the night.

Neon, the element that dances free,
In tubes and signs, for all to see,
Its radiance paints the urban scene,
A kaleidoscope of colors, serene.

From bars and clubs to bustling streets,
Neon's presence, it never retreats,
A beacon of life in the concrete maze,
Guiding lost souls through the haze.

In every corner, it claims its place,
Illuminating the darkest space,
Neon's glow, so bold and bright,
A symphony of fluorescent light.

Its hues of red, blue, and green,
Create a spectacle, a dazzling scene,
A neon city, alive and awake,
A vibrant dream, a visual earthquake.

So let us celebrate this noble gas,
And the beauty it brings as it comes to pass,
Neon, the element that shines so pure,
A testament to creativity's allure.

ELEVEN

ELEMENT GRAND

In the heart of the city, Neon resides,
A kaleidoscope of colors it provides.
With vibrant hues that catch the eye,
It illuminates the urban sky.

Neon signs, glowing bright,
Guide us through the dark of night.
A beacon of light, a symbol of hope,
In the concrete jungle, it helps us cope.

Its glow reflects on faces, creating a glow,
A mesmerizing aura wherever it goes.
A touch of magic, a flickering flame,
Neon's allure is never the same.

In the streets, it dances and twirls,
Unveiling secrets, illuminating worlds.
A symphony of light, a visual delight,
Neon's presence makes everything right.

So, let us celebrate this element grand,
With its radiant beauty, oh so grand.
Neon, oh Neon, we sing your praise,
Forever enchanting with your vibrant blaze.

TWELVE

CAPTIVATE THE SENSES

In the heart of the city, Neon dances in the night,
A symphony of colors, a shimmering delight.
It paints the urban landscapes, a vibrant hue,
A kaleidoscope of brilliance, so pure and true.
From the signs that flicker on the bustling streets,
To the glow of the bars where laughter meets,
Neon breathes life into the darkest alleys,
A beacon of enchantment, casting neon valleys.
Its vibrant shades, like a mesmerizing spell,
Captivate the senses, in a world they dwell.
From electric blues to fiery reds,
Neon's presence, like a dream, it spreads.
In the quiet corners where secrets hide,
Neon whispers softly, a luminous guide.

It illuminates the darkness, with a gentle grace,
A shimmering reminder of life's vibrant embrace.
  So let us raise our glasses to Neon's gleam,
A symbol of wonder, a radiant dream.
In the heart of the city, where life is aglow,
Neon's magic dances, forever to show.

THIRTEEN

SOUL OF THE NIGHT

In the heart of the city, where dreams come alive,
There's a magic that dances, where Neon does thrive.
A symphony of colors, a kaleidoscope of light,
Neon illuminates the darkness of the night.

With a flicker and a buzz, it captures our gaze,
A vibrant spectacle that never ceases to amaze.
In every glowing sign and shimmering display,
Neon paints the urban landscape in its own special way.

Oh, Neon, you're a beacon of creativity and art,
A symbol of inspiration that sets fire to the heart.
Your luminescent glow, so captivating and bright,
Ignites the spark of imagination, day and night.

You breathe life into alleys, and streets come alive,
With your electric presence, you make us thrive.

In every curve and letter, you whisper tales untold,
A symphony of dreams, in Neon's vibrant fold.

    Oh, Neon, you're a poet, painting stories in the air,
A mesmerizing language that all can share.
From the flickering signs to the buzzing cafes,
You cast a spell upon us, in your neon haze.

    So let us celebrate this element so rare,
Its beauty and enchantment, beyond compare.
Neon, the muse of artists, the soul of the night,
Forever in our hearts, your brilliance shines bright.

FOURTEEN

ELEMENT BRIGHT

In colors vibrant, Neon reigns,
A luminescent beauty, it sustains.
From atoms small, a brilliance born,
A light that dances, the night adorns.
   With hues that glow, a vivid display,
Neon's glow brings life, in its own way.
A noble gas, it shines so bright,
A beacon of enchantment, a radiant light.
   In tubes and signs, the city comes alive,
Neon's presence, it will never strive,
To fade away, or be forgotten,
Its glow persists, a vivid pattern.
   In streets aglow, the city's heart,
Neon's embrace, a work of art.
A symphony of colors, a dazzling show,
Neon's magic, it continues to grow.

It ignites imagination, sparks a dream,
A vibrant glow, a radiant gleam.
Neon, a symbol of wonder, it seems,
A testament to life's vibrant themes.
    So let us celebrate this element bright,
Neon's allure, a captivating sight.
For in its glow, we find delight,
A reminder to live with colors so bright.

FIFTEEN

DARKNESS RETREATS

Neon, the beacon of light,
In the darkness, shining bright.
A symbol of hope, it glows so clear,
Dispelling shadows, erasing fear.

A mesmerizing aura it possesses,
A captivating hue that never ceases.
With its vibrant shades, it paints the night,
A kaleidoscope of colors, a mesmerizing sight.

In the neon glow, the city awakes,
A vibrant tapestry, a dream it makes.
From the alleys dark to towering heights,
Neon illuminates, ignites.

It dances with the shadows, oh so bold,
A luminescent story yet untold.
In its presence, darkness retreats,
A symphony of light, it completes.

Oh, Neon, your radiance divine,
A touch of magic in every line.
Forever enchanting, forever bright,
You fill our world with pure delight.

SIXTEEN

KALEIDOSCOPE

In the heart of the city, Neon's glow prevails,
A mesmerizing force that never fails.
It paints the darkness with vibrant hues,
And brings to life what was once subdued.

Through the misty night, its light cascades,
Guiding lost souls on their wandering escapades.
In the neon-lit streets, dreams come alive,
Igniting passions that strive to thrive.

Neon's dance is a symphony of delight,
A kaleidoscope of colors that shine so bright.
It whispers secrets in the midnight air,
And lures you in with its enchanting flair.

Neon, a magician of the urban scene,
Transforming the mundane into something serene.
With a flicker of light, it sparks imagination,
Creating a vibrant atmosphere of elation.

So let Neon's glow be a beacon of hope,
A reminder that life is a kaleidoscope.
In the darkest corners, it'll always be found,
Bringing enchantment to the world around.

SEVENTEEN

DANCE OF ELECTRONS

In the deepest alleys, where shadows reside,
A force of enchantment, Neon does hide.
A flickering glow, a vibrant embrace,
It paints the darkness with colors and grace.

A dance of electrons, a mesmerizing sight,
Neon's brilliance illuminates the night.
A gas so noble, so captivatingly bright,
It sparks inspiration and ignites our delight.

A beacon of life, it whispers a tale,
Of dreams and desires, it never shall fail.
In the realm of chemistry, Neon does shine,
A symbol of wonder, so divine.

So let us marvel at its ethereal glow,
And let Neon's magic forever flow.

For in its presence, we find solace and cheer,
Neon, the element, forever held dear.

EIGHTEEN

VIBRANT HUE

Neon, oh neon, a captivating force,
In the city's veins, a powerful source.
Glowing in colors, vibrant and bright,
Guiding our way through the darkest of night.

A dance of electrons, so pure and rare,
You bring life to the city, a shimmering flare.
With your neon glow, you paint the town,
A kaleidoscope of lights, all around.

In the signs of the city, you come alive,
In every storefront and street, you strive.
Your brilliance enchants, a spectacle to behold,
Transforming the mundane into something bold.

Neon, oh neon, a symbol of dreams,
You light up the night with your vibrant beams.
From the bustling streets to the quietest lane,

You bring inspiration, breaking free from the mundane.

In the midnight hour, when all is still,
You guide lost souls, with an ethereal thrill.
A beacon of hope, shining through the haze,
Leading us forward, in these uncertain days.

Neon, oh neon, a flickering flame,
You illuminate the darkness, with no shame.
In your radiant glow, we find solace and peace,
A gentle reminder, that darkness will cease.

You spark inspiration, ignite the fire,
In the hearts of dreamers, you never tire.
Oh, neon, you're more than an element of science,
You're a symbol of art, and our deepest defiance.

So shine on, neon, in your vibrant hue,
In this ever-changing world, we turn to you.
For in your glow, we find a sense of delight,
A reminder that even in darkness, there's always light.

NINETEEN

RADIANT SPIRIT

In the heart of the city, Neon dances with delight,
A symphony of vibrant hues, a mesmerizing sight.
Like a celestial enchantress, it flickers and glows,
Bringing life to the darkness, wherever it flows.

A whisper of brilliance, a shimmering display,
Neon paints the night sky, in its own special way.
A kaleidoscope of colors, an electric embrace,
Igniting imagination, with its captivating grace.

Through the bustling streets, Neon comes alive,
Illuminating the shadows, helping dreams to thrive.
It transforms the mundane into a magical scene,
A beacon of hope, where inspiration convenes.

Neon weaves its spell, casting dreams in the air,
A symphony of radiance, beyond compare.
It dances with the stars, painting dreams in flight,
Guiding lost souls, with its ethereal light.

Oh, Neon, you're a muse, a wonder to behold,
A source of endless inspiration, pure and bold.
In your luminescent glow, the world comes alive,
With your radiant spirit, dreams will always thrive.

TWENTY

ESSENCE OF BEAUTY

In the darkest hour, a shimmering light,
Neon, the element, casting its glow so bright.
A gas of wonder, with electric might,
A hue that captures, an ethereal sight.

Through city streets, Neon's dance unfolds,
A vibrant tapestry, its story never told.
It paints the night with colors bold,
A symphony of brilliance, a sight to behold.

Neon, the guide for lost souls to find,
A beacon in the darkness, a gentle reminder.
It illuminates paths, ignites passions in kind,
Leading us forward, leaving shadows behind.

In mundane moments, it weaves its spell,
Transforming the ordinary, casting a spell.
From flickering signs to art that compels,
Neon brings beauty, where once none fell.

A symbol of hope, Neon's radiant glow,
In times of despair, it helps dreams grow.
With every flicker, it whispers, "you know,
You're not alone, let your spirit flow."

Defying the darkness, Neon takes flight,
It pierces the night, igniting our might.
In its luminescence, we find our sight,
A world of possibilities, shining so bright.

Neon, the artist's muse, a source of devotion,
Inspiring creations, with boundless emotion.
In its glow, we find art's truest notion,
A symbol of defiance, a vibrant commotion.

So let Neon's brilliance forever ignite,
A beacon of art, inspiration, and light.
In a world ever-changing, it shines so bright,
The essence of beauty, forever in our sight.

TWENTY-ONE

SHOW US THE WAY

Neon, oh Neon, radiant and bright,
A beacon of hope in the darkest night.
Your vibrant glow, a celestial dance,
Ignites imagination with every chance.

In the streets you flicker, a neon haze,
Transforming the mundane with your vibrant blaze.
A symphony of colors, a visual delight,
Guiding lost souls with your mesmerizing light.

You paint the city with your electric hue,
A kaleidoscope of dreams, both old and new.
Through bustling streets and empty alleys,
You whisper secrets to the night-time rallies.

Oh Neon, you defy the shadows' hold,
A symbol of defiance, so bold.
You spark creativity, ignite the flame,
In the hearts of artists, you etch your name.

In galleries and theaters, you take center stage,
With your luminescent glow, an eternal gauge.
You inspire poets, musicians, and all,
To break the boundaries and heed your call.

Neon, oh Neon, a symbol of art,
You illuminate souls, you captivate the heart.
In a world that can sometimes be dark and grey,
You bring forth light, and show us the way.

TWENTY-TWO

VIVID DISPLAY

Neon, the vibrant light that dances in the night,
A beacon of hope, pushing back the darkness with all its might.
Invisible to the naked eye, yet impossible to ignore,
You illuminate the world, leaving us in awe.

A gas so rare, but so full of life,
You defy the ordinary, cutting through the strife.
With colors so vivid, they ignite the soul,
Neon, you possess a magic we can't control.

In the signs of the city, you come alive,
Guiding lost souls, helping them survive.
Your glow, a lighthouse in the darkest hour,
Leading the way with your neon power.

You turn the mundane into a work of art,
Transforming the ordinary, igniting a spark.

Through your ethereal glow, creativity takes flight,
Inspiring dreams and visions, day and night.

    Oh, Neon, you are a symbol of defiance,
A reminder that even in darkness, there's a chance.
You bring joy and solace to the weary heart,
A burst of delight, a masterpiece of art.

    So shine on, Neon, in your vivid display,
Illuminate our world, chase the shadows away.
With your enchanting light, you cast a spell,
Neon, you are the magic we know so well.

TWENTY-THREE

ILLUMINATES THE SOUL

Neon, a gas of noble birth,
A light that shines upon the earth,
A hue that's bright yet soft and mild,
A color that's distinct and wild.

A beacon of hope that leads the way,
A source of inspiration every day,
A symbol of art that's bold and new,
A spark that ignites creativity in you.

It transforms the mundane into something grand,
And adds a touch of magic to every land,
A power that defies the darkness of night,
And brings forth beauty that's truly bright.

Neon, a force that illuminates the soul,
A light that makes us feel whole,

A color that sparks dreams anew,
A gas that's simply pure and true.

TWENTY-FOUR

GLOWING MASS

   Neon, oh neon, a gas so rare
Yet its glow is found everywhere
From signs on streets to colors in air
A symbol of art, it's beyond compare
   In the darkness, it shines so bright
A beacon of hope, a guiding light
It inspires, ignites, and takes flight
A source of creativity, a wondrous sight
   In science, it's a noble gas
But in art, it's a master class
A canvas of colors, a glowing mass
Neon, oh neon, it's truly first-class
   Its electrons, oh so unique
Their dance creates a stunning mystique
A transformative power, it does wreak
Neon, oh neon, it's beauty we seek

So let it shine, let it glow
Let its colors put on a show
Neon, oh neon, it's art, you know
A gas so rare, yet it steals the show.

TWENTY-FIVE

STORY OF SCIENCE

Neon, oh Neon, a noble gas so pure,
Invisible to the eye, yet so demure,
A force that illuminates the soul,
Inspiring creativity to take its toll.

A glowing light, so bright and bold,
A story of science waiting to be told,
Transforming matter with its spark,
A power so strong, it makes its mark.

In art, it finds its perfect place,
A canvas of color, a glowing face,
A transformative power, a brilliant hue,
Neon, oh Neon, we're in awe of you.

A beauty that captivates and charms,
A unique element with so much to disarm,
From science to art, you shine so bright,
Neon, oh Neon, you're a wondrous sight.

TWENTY-SIX

BRIGHTER DAY

In the realm of luminescent dreams,
Where darkness with brilliance gleams,
A gas of vibrant hue unseen,
Neon, the element supreme.

From the heavens it descends,
With a glow that never ends,
A cosmic dance of neon lights,
Guiding ships on starry nights.

In the city's bustling streets,
Where life and energy competes,
Neon signs come to life and shine,
A symphony of colors so fine.

A symbol of defiance it becomes,
Against the night, it boldly thrums,
With rebellious spirit, it ignites,
A beacon of hope in darkest nights.

In laboratories, it sparks new life,
Unveiling secrets, cutting through strife,
Transforming the invisible to reveal,
The wonders that science can unseal.

Neon, a muse for artists' delight,
With its brilliance, their canvases ignite,
A palette of colors, vivid and bold,
Inspiring tales that must be told.

In the heart, it stirs a fire,
A creative spark, a burning desire,
To break free from the mundane,
And let imagination reign.

Oh, Neon, you are a wondrous thing,
A catalyst for the songs we sing,
With your glow, you light the way,
Leading us to a brighter day.

So let us bask in your ethereal light,
And embrace the beauty you ignite,
Neon, the element of dreams,
Forever glowing, it seems.

TWENTY-SEVEN

MAGIC OF THE ATMOSPHERE

In the depths of the cosmos, Neon resides,
A shimmering marvel, where beauty coincides.
Its atomic dance, a symphony of light,
Igniting the heavens, painting the night.

Neon, the muse of artists and dreamers,
Ignites imagination, as it softly glimmers.
A palette of colors, vibrant and bold,
Unveiling stories yet to be told.

In the city streets, Neon finds its home,
Where creativity thrives, like a celestial dome.
Signs of enchantment, they come alive,
Guiding lost souls, helping them survive.

With a flicker of brilliance, Neon reveals,
The secrets of darkness, it gently conceals.

A beacon of hope, cutting through the haze,
To illuminate paths in mysterious ways.
    Neon, the alchemist of transformation,
Turning mundane into art, a divine creation.
Its luminescence, a symbol of life,
A reminder to embrace both joy and strife.
    So let us marvel at Neon's vibrant grace,
And let it inspire, in every time and place.
For in its glow, we find solace and cheer,
A testament to the magic of the atmosphere.

TWENTY-EIGHT

CAPTIVATING AND GRAND

In the darkest depths of cosmic space,
A luminescent beauty takes its place.
Neon, the element of ethereal glow,
With brilliance that captures hearts below.

A gas so rare, yet captivatingly bright,
It dances with colors, enchanting the night.
A beacon of hope, it pierces the gloom,
Guiding lost souls to their destined bloom.

Neon, the muse of artists and dreamers,
Whispering secrets through vibrant gleamers.
It paints the canvas of the starry sky,
Igniting imagination, soaring high.

In the signs of the city, it comes alive,
Flashing and flickering, a neon drive.

Words and symbols glowing with desire,
Inspiring passions that never tire.
    With a hue that's pure and full of grace,
Neon illuminates the human race.
A catalyst for art, it sparks the fire,
Transforming the mundane into something higher.
    Oh, Neon, you hold the power to amaze,
In your radiant glow, the soul finds its maze.
A mystical force, captivating and grand,
A shimmering light that we'll forever understand.

TWENTY-NINE

RADIANT GLOW

Neon, the radiant light that fills the night,
A shimmering glow, a celestial delight.
In noble gas form, you dance with grace,
Illuminating the darkness, a cosmic embrace.

Your neon signs, a vibrant display,
Guiding lost souls along their way.
Through city streets, your colors ignite,
A kaleidoscope of brilliance, a captivating sight.

In laboratories, you reveal truths unseen,
A beacon of knowledge, a scientist's dream.
With your glowing hue, you unravel the unknown,
Unlocking mysteries, making the universe known.

Neon, you inspire artists to create,
With brushes and palettes, they emulate,
Your electric glow, a muse for their art,
A symphony of colors, a masterpiece's start.

From the canvas to the silver screen,
You paint stories, like a vivid dream.
In movies and theater, you set the stage,
A mesmerizing presence, an actor's cage.

Oh, Neon, you symbolize so much more,
Than a simple element, that's for sure.
You represent transformation and change,
A catalyst for progress, a world rearranged.

So let us bask in your radiant glow,
And embrace the magic you bestow.
In science and art, you'll forever remain,
Neon, the element that will never wane.

THIRTY

MYSTERIOUS WAYS

Neon, oh Neon, a gas so rare,
Inert and colorless, yet beyond compare.
It shimmers and shines with a hue so bright,
A sight to behold, a true delight.

It's not just a gas, it's so much more,
A symbol of progress, a key to explore.
Its transformative power, it can't be denied,
Bringing new life, where once it had died.

Neon, oh Neon, a muse to the arts,
An inspiration to all, it plays its part.
It illuminates paths in mysterious ways,
Igniting creativity, with its radiant blaze.

It transforms the mundane, into something higher,
A work of art, that sets hearts on fire.
It brings stories to life, with a neon glow,
A symbol of hope, a chance to grow.

Neon, oh Neon, a gas so rare,
A force of nature, beyond compare.
Its significance, we can't ignore,
A symbol of transformation, forever more.

THIRTY-ONE

LUMINESCENT DREAM

Neon, the radiant muse of the night,
A shimmering dance of colorful light.
In the darkness it comes alive,
Transforming the ordinary, making it thrive.

A beacon of hope, a symbol so bright,
Guiding lost souls with its celestial light.
Through city streets, it paints a new scene,
A kaleidoscope of colors, oh so serene.

In art and science, Neon takes its place,
A catalyst for innovation and grace.
From glowing signs to noble gases,
Its presence leaves us in awe, as time passes.

With every flicker and every glow,
Neon sparks creativity, causing hearts to grow.

Igniting inspiration, like a fire in the soul,
It breathes life into dreams, making them whole.
    Oh Neon, you're more than just an element,
You're a symbol of progress, a testament.
To the power of transformation, the beauty of change,
You bring light to the world, in a way so strange.
    So let us embrace this luminescent dream,
Where Neon reigns, casting its gleam.
A reminder that in darkness, there's always a spark,
And with Neon's glow, we'll illuminate the dark.

THIRTY-TWO

HEARTS OF CREATORS

In the dark of night, Neon shines bright,
A beacon of brilliance, a radiant light.
It dances in tubes, vibrant and bold,
A spectacle of colors, a story untold.
   Neon, the muse of artists' dreams,
Inspiring strokes, unveiling themes.
With brush in hand, they capture its glow,
Creating masterpieces that come to life and grow.
   From city streets to galleries grand,
Neon's allure, a sight so grand.
It sparks imagination, ignites the fire,
In the hearts of creators, it never tires.
   Oh, Neon, you are a magnificent sight,
A symbol of creativity, burning so bright.
Through your glow, stories come alive,
A testament to the magic you derive.

So let us celebrate this element true,
For it brings inspiration to me and you.
Neon, the muse, forever we'll adore,
For its brilliance and beauty, we'll forever explore.

THIRTY-THREE

TURNING PAGE

    Neon, oh Neon, a gas so rare
Inert and noble, without a care
A hue of red, a tint of blue
A muse for artists, a dreamer's view
    In tubes and signs, it lights up the night
A beacon of hope, a guiding light
In movies and plays, it sets the stage
A catalyst for progress, a turning page
    In science and art, it transforms the plain
A spectrum of colors, a neon refrain
A symbol of progress, a futuristic dream
A light in the darkness, a radiant beam
    Neon, oh Neon, a gas so pure
A transformative power, forevermore.

THIRTY-FOUR

CELESTIAL LIGHT

Neon, luminescent muse of the night,
A dance of colors, a celestial light.
In the city's heart, you shimmer and glow,
A tapestry of dreams, an ethereal show.

Artists find solace in your vibrant hue,
Brushstrokes of brilliance, their canvas anew.
You ignite their imagination's fire,
A catalyst for creation, their hearts aspire.

In the darkened streets, you guide the way,
A beacon of hope, a guiding ray.
Your neon signs, a symphony of design,
Captivating eyes, a mesmerizing shrine.

A cityscape transformed by your embrace,
A wonderland of hues, a vibrant chase.
In every corner, a story to be told,
Neon, your presence, forever bold.

So let us celebrate your radiant glow,
The beauty you bring, a spectacle to show.
Neon, the element that illuminates the night,
A muse for artists, a catalyst for light.

THIRTY-FIVE

CITY STREETS TO OPEN SKIES

In the realm where dreams ignite,
Neon dances in the night,
An element that's full of might,
A beacon in the dark, so bright.

With a flicker and a glow,
Neon paints a vibrant show,
In colors that we've yet to know,
A kaleidoscope that steals the show.

From city streets to open skies,
Neon's magic mesmerizes,
It whispers secrets, tells no lies,
A cosmic force that never dies.

In signs and art, it finds its place,
A muse for those with vivid grace,

Inspiring minds to reach the space,
Where imagination finds its embrace.
   So let Neon guide us on our way,
Illuminate the night and day,
With colors bold that never sway,
Forever sparking hearts, we say.

THIRTY-SIX

PROGRESS AND MIGHT

In the realm of vivid hues, Neon takes its flight,
A glowing ember in the heart of the night.
With atoms dancing, electric and bright,
It kindles dreams and ignites the mind's delight.

Neon, the luminary of the city's domain,
A beacon of hope, a radiant refrain.
From bustling streets to theaters' grand stage,
It breathes life into stories, page after page.

Artists, inspired by its vibrant glow,
Craft works of wonder, their talents aglow.
Brush strokes dance, as colors come alive,
Neon's transformative power, forever they'll strive.

With flickering signs and words ablaze,
Neon whispers secrets, in mysterious ways.

It guides lost souls, like a celestial guide,
Through the darkness, its luminosity beside.
    Neon, a symbol of progress and might,
A testament to human endeavor's height.
In laboratories, its secrets unfurled,
Unleashing the power that can change the world.
    So let us bask in Neon's radiant light,
A catalyst for dreams, burning ever so bright.
For in its glow, new stories are born,
And the world is forever transformed.

THIRTY-SEVEN

PAINTS THE NIGHT

In the dark, a glow appears, so bright,
A shimmering light that ignites the night.
Neon, the element, so noble and rare,
With brilliance, it captures the heart's stare.

A gas, it dances in glass tubes of art,
Creating a spectacle, a masterpiece's start.
Its vibrant hues, a palette so bold,
A symphony of colors, a story to be told.

Neon, the muse of artists and dreamers,
Whispering secrets, inspiring creators.
With every stroke and every brush,
It brings to life a world so lush.

In the city streets, it paints the night,
Guiding lost souls with its gentle light.
A beacon of hope, a symbol of progress,
Neon, the element, we are truly blessed.

So let it shine, let it glow,
Neon, our ally, our muse, our show.
For in its presence, we find delight,
A mesmerizing beauty, forever in sight.

ABOUT THE AUTHOR

Walter the Educator is one of the pseudonyms for Walter Anderson. Formally educated in Chemistry, Business, and Education, he is an educator, an author, a diverse entrepreneur, and he is the son of a disabled war veteran. "Walter the Educator" shares his time between educating and creating. He holds interests and owns several creative projects that entertain, enlighten, enhance, and educate, hoping to inspire and motivate you.

> Follow, find new works, and stay up to date
> with Walter the Educator™
> at WaltertheEducator.com

www.ingramcontent.com/pod-product-compliance
Lightning Source LLC
LaVergne TN
LVHW051958060526
838201LV00059B/3717